CANDLES IN FLOWER ARRANGEMENTS

CANDLES IN FLOWER ARRANGEMENTS

by Myrtis N. Powell

PHOTOGRAPHS BY ERNEST HARDIN

VAN NOSTRAND REINHOLD COMPANY
New York Cincinnati London Toronto Melbourne

Van Nostrand Reinhold Company Regional Offices:
New York Cincinnati Chicago Millbrae Dallas.

Van Nostrand Reinhold Company International Offices:
London Toronto Melbourne.

Library of Congress Catalog Card Number: 77 90829.

Printed in Japan.

Published in 1970 by Van Nostrand Reinhold Company,
A Division of Litton Educational Publishing, Inc.
450 West 33rd Street, New York, N.Y. 10001.

Published simultaneously in Canada by
Van Nostrand Reinhold Ltd.

16 15 14 13 12 11 10 9 8 7 6 5 4 3 2

ACKNOWLEDGMENTS

I am very grateful to my understanding family for their patience and cooperation while I was compiling this book—to my husband, who helped me carry the numerous arrangements to the studio for the pictures, and also helped me with the manuscript; to my children, for their interest; to my dear sister, Lassie Steadman, for all she has done for me. And particularly to my sister-in-law, Rosemary Powell, who accompanied me on so many trips to gather materials for the arrangements.

With special thanks

To my employers, Mr. and Mrs. Howard Johnson, for the use of the materials of their shop and for their time and advice.

To my photographer, Ernest Hardin, for his patience and skill.

To Gertie and Jack Truss for sharing their abundance of beautiful flowers, fruits, and vegetables.

To Margaret Hosford, Colonel George Kinzie, and Mary Wilson for some of the candles used in the arrangements illustrated in the book.

To Edward Barrera of the Candle Institute for advice and help on the chapter on candlemaking.

To Fanny Richardson, Lila Stovall, Le Claire Dewey, and Carol Lewis for their encouragement and praise.

And to those who spent long hours typing my text.

CONTENTS

1

THRESHOLD TO A NEW ART

Flower arranging is as much an art as painting. It uses the same elements and principles—although of course a good flower arranger, instead of merely depicting the beauties of nature, uses them in such a way that others can see and enjoy them directly. Candlemaking is also an art; it is essentially a kind of sculpture.

Combining these two arts presents a new challenge to both the arranger and the candlemaker. From the arranger's point of view, candles may be used in such a way that they will complement the arrangement. Candles can strengthen the arrangement and add distinction, or they can be the dominant part of the arrangement, or they can lend support to the plant material. From the candlemaker's point of view, flower arranging enormously expands the decorative possibilities of candles. Very often people come to me for advice on how to do more with beautiful candles they have bought or made, and I hope this book will be their answer.

Perhaps you think of candles as primarily functional. They were, of course, first created simply to produce light. But that the candle brought more than light is evidenced by the refusal of civilized society to give it up entirely for any modern lighting system. This has been true because candlelight, when properly used, stimulates the inner feelings of people and adjusts their emotions to the occasion. It occurred to me that the combination of such an influence with the ancient art of flower arranging could be more effective than either candles or flower arrangements are alone, and I think I have proved I was right. I hope I can encourage the flower arranger to make or buy candles for arrangements and inspire the candlemaker to try flower arranging.

Many people are not sure when it is appropriate to light candles. There are very few hard-and-fast rules, but the following are a few hints:

1. Usually candles are not used burning on a luncheon table, but it is permissible when they are needed for light on a dark day. Never draw draperies to darken the room for lighted candles.

2. Lighted candles are fine on a tea table, provided the tea starts after two thirty in the afternoon.

3. Always light candles at an evening meal or reception.

4. On special occasions such as Christmas and Thanksgiving, when candles are a part of the decoration, they can be lighted regardless of the time of day.

5. Lighted candles are always correct at any function for a bride, again regardless of the time of day.

The most common use of lighted candles is on a dinner table, and here are a few hints about that:

1. Avoid candle flame at eye level.

2. Candles used outdoors should have some protection from the wind.

3. The heights of stemware, candles, and flowers should be different to avoid monotony.

4. Candelabra can be used to contain flowers as well as candles to conserve space on table.

5. When selecting candles for table, texture, size, and color should be considered. Use candles of one color, not a variety of colors.

6. If candles are refrigerated a few hours they will burn more slowly.

You can buy ordinary candles or, in specialty stores, a variety of handmade ones. It is wise, even if you make your own, to keep a small supply of run-of-the-mill candles on hand. Any candle can be easily decorated. Frequently when time is very limited plain candles save the day. They can be decorated in many ways with whipped wax, jewels, wax flowers made from sculpturing wax, dried or artificial flowers and fruits, old beads, braids, and so on. If you do not have just the right shade of candle, try rubbing the candle with colored chalk and then polishing it with a soft cloth. Plain candles can also be sprayed with florist spray for a particular color effect. However, neither method of coloring achieves the same attractive glossy color that can be produced at the time the candle is made. Florist spray serves well when time is an element in preparation, or when one desires the particular effect of the sprayed-on color. Spraying candles with pearl spray gives an elegant pearly finish—in fact, sometimes an entire arrangement is so sprayed to produce a finish that is very effective for some uses.

It is usually advisable to purchase all tapers commercially. You can achieve a hand-dipped effect by dipping the commercial taper once in hot wax. (There may be times when you want the handmade taper for its intrinsic value and feel that the commercial taper is inappropriate, but remember that making small tapers requires almost as much time and is much more tedious than making large pillar candles.) Dripping white tapers lightly with colored wax to harmonize with a color combination is very effective and makes them very different from the plain dinner tapers. This decoration is particularly good in candelabrum arrangements and for special occasions such as teas or wedding receptions.

Of course you can do a great deal with commercial candles—they are used in many of the arrangements in this book. But I hope that you will advance to making your own. I found I was limited by the shapes and colors of candles that could be purchased. I then turned to making candles to complement a particular container or special arrangement I had in mind. The basics of candlemaking, which are explained in the last chapter in this book, can be mastered in a few evenings. If you know all about candles but not very much about flowers and foliage, do not be afraid to learn—the basics are simple.

Fresh plant material, which is much easier to use than dried material or artificial material combines very well with candles, especially for holidays and parties. The colors and textures of fresh plant material cannot be duplicated in artificial materials. It is very important—but very easy—to condition fresh (or live) flowers and foliage: Cut the ends of the stems on a slant early in the morning or in late afternoon, put the stems in warm water immediately, and leave them in the water for at least two hours. They absorb the water and this preserves the flowers. Foliage or leaves to be used in arrangements should be completely submerged in water for at least two hours. Some flowers require special treatment. Flowers with woody stems should have the ends of their stems crushed before they are placed in the warm water. Stems that contain a milky substance "bleed" when cut, so they require a slight searing over a candle flame or other heat to seal the pores. Some flowers such as tulips have limber or fragile stems and it is wise to wrap them in a water-absorbing paper up to the flower heads before placing them in water to protect them from damage. These simple precautions will lengthen the life of flowers used in arrangements.

What has been said above about dried materials is not meant to imply that they have no place in arrangements with candles. They have a very definite place, and they add variety. Some dried or treated materials that go well with candles are magnolia foliage, straw flowers, tropical foliage of palms, and seed pods, and there are many others. The dried plant material is not as pliable as live material, and it must be handled carefully to prevent breakage. Dried foliage can be treated with a glycerin-and-water solution to lessen its crispness, but when this is done most materials change their colors to tints and tones of brown. You can use your imagination and invent your own dried materials One of the most attractive candle arrangements I have includes three bunches of "grapes" that I made from peach pits. (A photograph of this arrangement appears in Chapter III.)

The third type of material that combines well with candles is made of polyethylene or other synthetic substances. Many people prefer artificial materials for candle arrangements because they are more permanent than natural ones. There have been improvements in recent years; colors often come very close to the natural colors of live flowers and foliage. Bubble branches, shown in several of my arrangements, combine dry and artificial materials—they are bare branches covered with styrofoam balls. You can make your own: Buy a bag of styrofoam bubbles, put them in boiling water, and swish the branches in the water.

Each kind of material has its place in flower arranging and can be used with candles to create successful design—which is the subject of the next chapter.

2

DESIGN

Design in flower arrangements is the combination of five elements: line, form, texture, color, and pattern. Candles can help supply any or all of these elements in a design. Let us see how this is done.

Line is the skeleton of a design. Candles can lengthen the design in height and even in width, making the emotional feeling stronger. (The vertical line gives a feeling of alertness, the horizontal a feeling of relaxation.)

Form is essentially the shape of an arrangement. Candles can determine the shape of a design themselves or they can support the shape of the design of the plant material.

Texture is the surface quality of the material used, whether smooth or rough, shiny or dull, coarse or fine. Texture gives sparkle to flower arrangements and brings them to life. Candles can add distinction to a flower design by presenting another surface quality—they can have textures that plants cannot. However, not too many different textures should be used in an arrangement or the effect will be weakened.

Color is the most dangerous element in design, and at the same time it is the most important. Candles can provide the most vivid color within the harmony of a design, or they can merely complement the colors of the plant material. Warm colors such as red, orange, and yellow advance, and therefore they should be used low and usually to the center. Cool colors such as blue, violet, and green recede, so they are used high and to the sides of an arrangement. The colors in an arrangement must be positioned so that no one part seems to leap out toward you.

Pattern, like form, is concerned with the shape of an arrangement, but in terms not just of silhouette but of solids and voids, or space between materials. Variety in sizes and forms of material and sizes and forms of voids makes an arrangement more interesting and gives it more distinction. Candles can be very useful in creating a pattern—unlike plant materials, they can have any shape you like and so provide an unlimited variety of solids and voids.

A good arrangement, one that combines the five design elements successfully, will exemplify six principles. These are balance, dominance, contrast, rhythm, proportion, and scale.

Balance is stability—a feeling that the arrangement will not tip over. A candle might be the one element that will add balance. Balance must be considered in relation to the space the arrangement occupies—balance between the plant material and the container as well as all other relations. A balanced design can be symmetrical or asymmetrical. A symmetrical design has the same amount of material on either side if a vertical line is drawn through its center; an asymmetrical design has unequal amounts of material on either side. We can obtain balance by the use of different materials, colors, lines, and textures. The candle can supply any one or all of these elements, completing the balance of a flower arrangement.

Dominance is the relative importance of one of the elements in a design. Obviously there is more than one color, one line, one texture, but one element is always more dominant in a good arrangement. A candle can dominate by its color, line, or texture, or it can merely lend support to the plant material.

Contrast is variety in the arrangement; it adds spice and emphasizes differences in line, color, materials, and textures. Candles can certainly provide any of these in a design.

Rhythm is movement in a design. It is obtained by gradation of sizes, colors, and lines. Arrangements are dead without rhythm. Candles can impart rhythm by continuation of the color, size, or line of an arrangement.

Proportion refers to the relative amounts of each material in the container or the amount of materials compared to the size of the container. The candle can give the needed proportion to a heavy container, as plant materials often cannot.

Scale deals with size only—the relation of the arrangement to the place in which it is to be used, the size of flower to the size of container, the comparative sizes of different plant materials, and so on. The size of the candle is an important consideration. Size and outside finish of the candle must harmonize. A large heavy candle would not be used with a fragile material.

Containers play an important part in design. Candles can be fixed in most containers that are used in flower arrangements. They can be anchored to oasis, floral clay, or needlepoint holders in order to keep them secure. If oasis or needlepoint holders are used, heat an icepick, make a hole in the bottom of the candle, and insert a wooden florist pick; this is thrust into the oasis or needlepoint. (Oasis is a spongy material that soaks up water and keeps plant material fresh. Be sure the oasis is saturated with water when the arrangement is completed.)

Containers are very important to any type of arrangement. The container should be in harmony with the plant material in color, texture, and shape. A candle can do much to balance a heavy container and plant material. However, a candle should not overpower the container and make the arrangement look top-heavy.

Finer or more delicate containers require the more delicate flowers and candles. Silver and china or crystal containers are widely used for arrangements for dinners, teas, and more formal functions. Brass and copper and similar materials call for bright plants. Bronze is an excellent material, because of its earthy color.

Try fruit and vegetables—such as melons, pumpkins, and squashes cut in half—for containers to be used on the patio and in breakfast rooms and dens. The cut fruit may be wiped dry with a paper towel, then rubbed with lemon juice to prevent it from turning dark; or it may be sprayed with a lacquer (hair spray) to preserve the natural color. Fruit so treated will usually last several days. Driftwood is also excellent for den and patio arrangements, and candles combine very well with wood. Small candleholders are available that screw into the wood. Usually materials found on the beach and in the woods harmonize with driftwood.

Most people overlook the best source of unusual containers: the junk pile.

Old lamp bases, plow discs, coal-stove grates...many, many pieces of scrap invite the eye, and I have used a number of such "found objects" in the arrangements pictured in this book.

Once you have learned something about the theory of design, you will want to get some materials together and try to assemble an arrangement. There are three types of design an arrangement can have: line, mass-line, or mass. A line arrangement is a design in which line is the dominant part. The line should be obviously an important part of the arrangement, and it should be attractive to the eye. It is characterized by restraint of material and openness of silhouette. Modern designs are apt to be line arrangements. Mass-line design has light mass added. Mass-line arrangements have height, width, and depth; they are most adaptable and most popular for use in American homes. Mass design has closed form; the design is full, but never crowded. It has depth, depending upon quantities of plant material and the grouping of flowers according to their kind, texture, and color. It has a linear pattern and may be symmetrically or asymmetrically balanced. This type of design is widely used with period furniture. Good massed designs are more difficult to make than line or mass-line arrangements.

Dinner tapers and slim tapers are better to use in line arrangements and line-mass arrangements. A thin taper, a few bare curved vines, and a few flowers can furnish ample material. The mass arrangement usually combines an abundance of plant material with small thin tapers; several are illustrated in this book. The large pillar candle may be used in a mass or mass-line design. Frequently the candle is the most dominant material as well as the most dominant texture.

The lines of the plant material or the candle should set the pattern for the type of the design. Consideration should be given to the place the design is to be used. Candles can give balance by their position in the arrangement. Always be sure the candles used are needed to complete the design, never adding a candle just because it has a pretty color or unusual shape, any more than you would add an extra flower because you have one left over.

Use the taller thinner material at the top of the arrangement, gradually placing the larger bolder forms and colors at the bottom of the arrangement. Interesting foliages should be used for lines and gradation of form. Use a variety of colors and textures, always grouping colors—warm colors (red, yellow, and orange) are used low in the arrangement, cool colors (green, blue, lavender) are used at the top and sides. Make a gradation of lines, forms, and colors to obtain rhythm.

All good designs should have good balance. When the arrangement is finished stand ten feet away and ask yourself, "Does this design look top-heavy?" If so, try adding a large leaf or open flower or perhaps a small stone, a candle, or some other material that will give a feeling of weight. Check to see if the container, the candle, and the plant material make a complete unit. Remember, if the finished design pleases the arranger it is successful—enjoy it.

Don't be afraid to be different! Distinction is gained by using everyday materials, containers, and candles in an unusual combination.

The black-and-white pictures that follow are meant to illustrate the part candles can play in the design of flower arrangements. (I have usually indicated what the colors in these arrangements are, for those who have good imaginations.) The color pictures in later chapters will also be instructive, although the drama of color makes some elements of a design less apparent and can even disguise a poor arrangement.

The tall earth-tone candle decorated with dripped wax forms the dominant feature of this design. The bunches of peach-pit "grapes" add variety in form and texture. The container is the top of an old coal heater inverted and mounted on an electric-motor housing. Distinction was achieved by combining waste materials with a giant glossy candle.

This mass arrangement illustrates the use of the candle as a subordinate element of the design. The plant material dominates the container and candle. Color relationship exists between the container of earth-toned bronze and the light-brown candle. The magnolia blossoms and shiny leaves complete the harmony of the design.

Exquisite textures of the roses and glass emphasize the smooth glossy finish of the candle in this arrangement. The broken brandy snifter was an accident that was turned into a conversation piece.

Part of an old weathered tree trunk and roots form the container, which dominates the design. The wood is adorned with plant materials that might be found in the trunk's natural surroundings—violets and ferns. These are accented by the pale-green candles, which contrast in color and texture with the other materials.

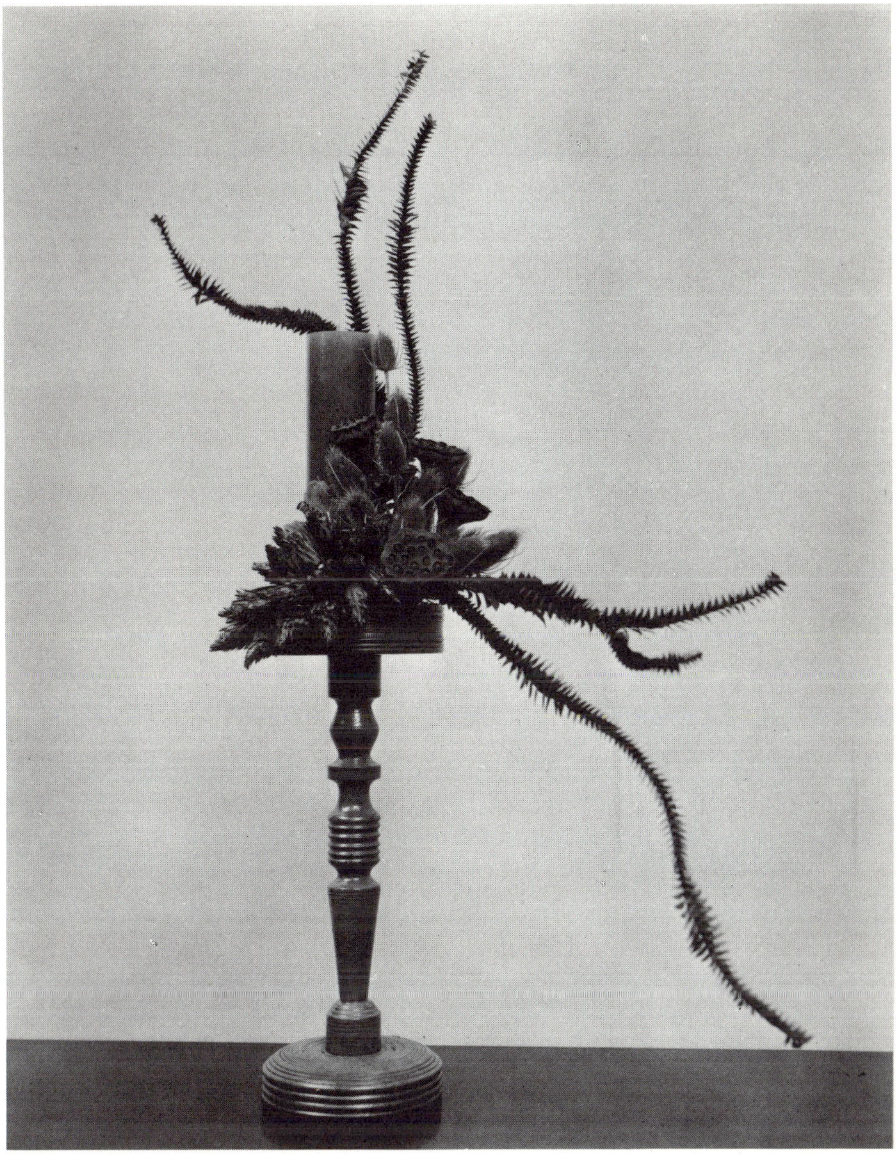

The smooth lines of the floor candleholder and the candle are accented by the zigzag lines of the bunya bunya branches. To complete this hearth floor arrangement, lotus pods and thistle and a giant bunch of grapes made from seed pods are used. All the colors are muted tones of green and brown

The candles in this vertical arrangement add height to the plant material and also continue the rhythm of the clipped fern and fishtail palm, which, like the candles, are chartreuse. A bunch of grapes emphasizes the contrast of the green candles and foliage and the black candleholder.

The hurricane candles in this horizontal arrangement are the most dominant element. The oval base was selected to emphasize the square shape of the candles. The natural plant materials, cones and evergreen foliage, are in thematic harmony with the animal scenes on the candles.

The round pillar candle was chosen to harmonize with the candleholder, made from a carved antique table leg. The forms of the fern, fronds, fishtail palm, and yarrow heads add the spice to this arrangement so necessary to good design. The candle adds strength to the color combination of brown tones.

This asymmetrically balanced arrangement features the pillar candle used in an openwork iron container. The heavy container and candle require the lighter plant material of wild dock seed spikes and the velvet texture of the roses. The design is in pleasing shades of green and yellow.

This design might be named "A Ship in the Storm." The dried palm was shaped in swirling curves to accent the low candle, which is red and represents the lighthouse. The three small milkweed pods could very well be the lifeboats. The forms and lines in the arrangement add distinction.

The slim square tapers are very necessary to this candelabrum design. They add height, which results in good balance. The white cupid candelabrum is decorated for a tea, using flowers and foliage in green and yellow to harmonize with the green tapers. I used white daisies, white Queen Anne's lace, and spikes of the golden rain tree.

The dominant piece of weathered wood used in a vertical position mounted on a wood burl needs little to complete the design. A red candle is used to add contrast of color and texture. The gladiola sprays, also red, complete the harmony and rhythm of color.

Butterfly weed and butterflies lend variety in form and color. Hosta leaves are used to add weight and transition of forms for perfect rhythm between candle and flowers. The leaves supply graduation of sizes in the design

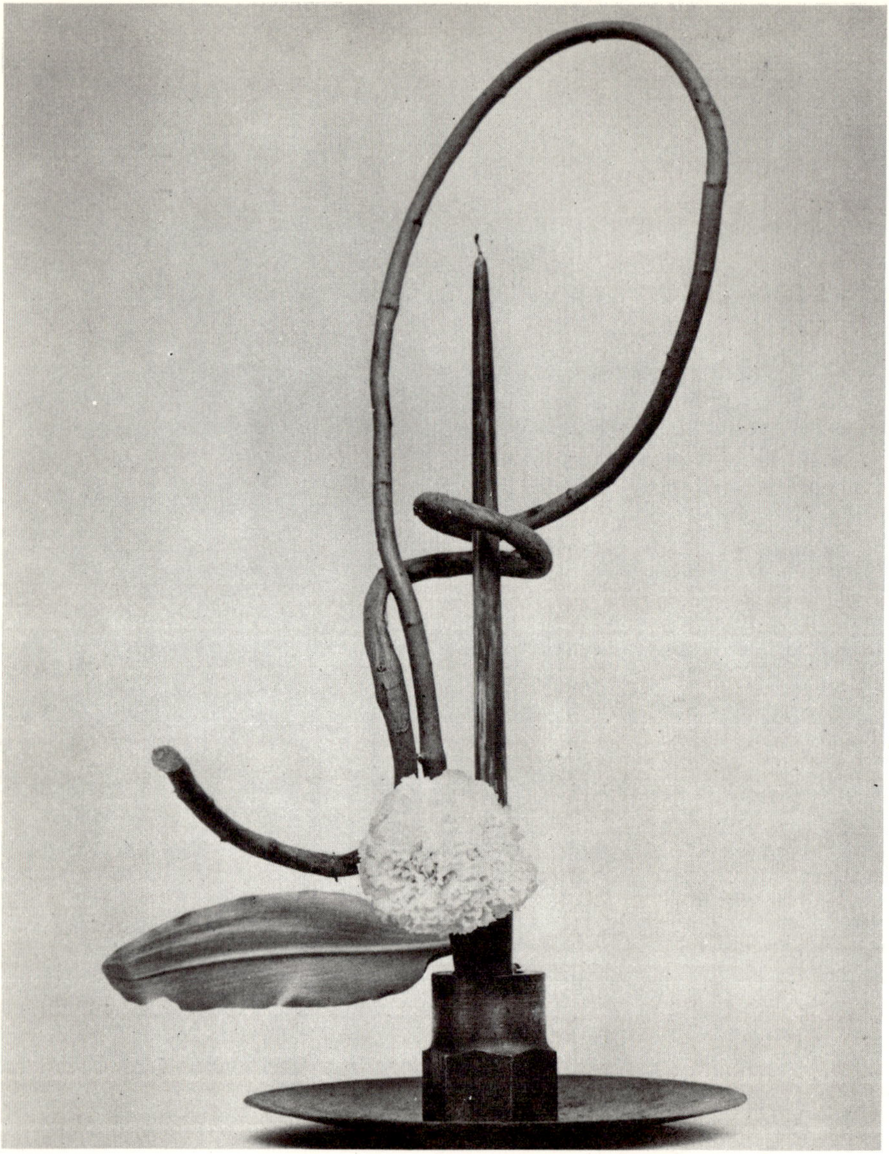

This modern design incorporates a candle to give the feeling of an abstract arrangement. The twisted wisteria vines, a leaf, and one blossom are fixed in a metal disk, using an oversized nut for a holder. This arrangement was designed by Mrs. Earl Stamps.

An arrangement of magnolias and candles, suitable for a patio party. It is a mass design, asymmetrically balanced, stressing the harmonizing textures of the velvety magnolia blossoms and the long glossy leaves. The container is a lazy susan, and a small tiered candelabrum holds the hurricane globe. The candles are small twisted tapers wrapped in honeycomb beeswax sheets. Very Southern and very fragrant.

The brass candle tree holds yellow and orange square taper candles. The strong vertical lines are softened by polyethylene grapevines entwined around the candle tree branches. This was used at a wedding reception, but the candle tree is just as beautiful at Christmas or, with ivy and flowers, as a church decoration for a wedding.

This Easter arrangement uses the unusual calla lilies in a crescent-shaped line to adorn the beautiful brown Madonna. The small candle at the base of the Madonna reflects the light upward, creating unusual shadows on the design. As is customary, the Madonna candle has no wick; it is not meant to be lighted.

In this Christmas design the star candle is used in a giant hurricane globe. Gold-glittered cattails and fruit emphasize the elegance of the materials used. Within the green, white, and brown color scheme, the spruce foliage contrasts with the cones and the white bisque angels complement the white carnations.

The lovely Madonna and Child figurine needs little additional material. The long narrow leaves form a background. Bases of two sizes are used for stability, and two small votive candles covered with crushed glass complete the balance and add variety of form and texture.

Notice that in black and white, this arrangement does not seem unified. But the green candle is in complete harmony with the green lichens at the base. Cryptomeria foliage and red holly berries furnish the plant material. A small elf dressed in red sitting on a toadstool balances the asymmetrical design and adds variety.

This arrangement illustrates variety of forms. The palmetto palm, sprayed red and clipped to add distinction, contrasts in form and color with the two block or pillar candles, which, like the container and grapes, are chartreuse. The grapes are wax, made by using a melon-ball scoop to shape soft wax; a wire is inserted into each grape.

Very little adornment is needed to create the delicate arrangement of Madonna and angels. Angel hair is used in the brandy snifters, and at the top is a small twisted candle—seemingly a small part of the design but a very important one.

3

CANDLE ARRANGEMENTS THROUGHOUT THE YEAR

In the past it was thought that candles should be used only at holiday seasons or at weddings and parties. But candle arrangements can be enjoyed throughout the year, especially during the spring, summer, and fall, when there is an abundance of flowers blooming in the garden. Flowers are not essential for a lovely candle design. Colored foliage—leaves such as caladium, begonia, or any pretty leaves—can be combined with candles for unusual designs. In winter, evergreen foliages, needled and broadleaf, make lovely decorations that last for weeks. Foliages offer a variety of forms and textures essential to a good design. Candles may be harmonized or contrasted in color, texture, and form with the other materials used in the arrangement. Dyed foliage, wild grasses, branches, or vines can be made into an arrangement, in tones of brown and beige, if they are treated by putting a solution of one part glycerine and two parts water into the container that holds the foliage design. The entire arrangement will thus be preserved. The material will cure and stay soft and pliable. Magnolia leaves, oak leaves, and any foliage that does not shed easily can be preserved in this manner, and the brown foliage can be combined with fresh blossoms. Daffodils and treated material combine well for hurry-up arrangements. Slip the fresh blossoms in with a preserved arrangement and you have a lovely combination fit to grace any table.

If fresh flowers are not available—or even if they are—fruits and vegetables are still another challenge to the candle and flower arranger. A few points should be remembered. If you use a number of fruits or vegetables, you might be tempted to include too many of similar shape—several round fruits such as apples, cantaloupes, and oranges, for example. To avoid monotony, try cutting one fruit or vegetable in half and placing it in the design—perhaps a bunch of grapes spilling out of a halved cantaloupe. (Do not forget to dry the cut fruit with a paper towel and either rub it with lemon juice or spray it with hair lacquer to keep it from discoloring.) Foliage can be used to tie the parts of the arrangement together or to complete the unit of fruits. Avoid too many contrasting colors of vegetables or fruits in any one design. Fruits and vegetables offer great variety in textures. The smooth surfaces of tomatoes, bananas, and grapes accentuate the roughness of artichokes and pineapples and add spice to the arrangement. Wax fruit, with its smooth texture, combines with the rough or smooth quality of the fresh fruit or vegetables. Foliage of vegetables offers as wide a variety as the foliage of shrubs and flowers. Beet leaves and stems are a shade of deep reddish purple, rhubarb stems are pink. Curly-leaf mustard and green onion tops are natural foliages for a fruit and vegetable arrangement.

Driftwood with candles and plant material makes distinctive arrangements for a den, patio, or playroom. Use a combination of materials that harmonize with the naturalness of the driftwood, like a combination of ferns, candles, birds, and toadstools.

Beautiful flowers and candles always harmonize. The type, color, and shape of the flower will determine the choice of the candle. Color harmony in the same tones and tints may be desired. If this is done one should usually select the candle in a lighter tint than the flower if the candle is to be a subordinate part of the arrangement, and in a deeper tint if it is to be the dominant part. The plant material should complement the candle.

The color pictures that follow show a variety of arrangements. Some of them are particularly appropriate for certain seasons or occasions, but none has so definite a thematic content that it is restricted to a particular use.

This floor candle arrangement, 36 inches tall, is beautiful when used on a hearth. An old ashtray stand antiqued in green and brown forms the base. Polyethylene grapevines and a profusion of grapes—including peach pit imitations for variety of texture—add to the mass and attractiveness of the tall dripped pillar candle.

A tall pewter-washed copper compote provides the container for an arrangement in green and yellow. Long leaves from the Peruvian daffodil are used to form the linear pattern, but iris foliage would do equally well. Two lemon candles with chartreuse pampas grass and a bunch of grapes complete the design. A typical arrangement without flowers.

Another candle arrangement without flowers. The rex begonia leaves add a variety of color, texture, and form to the prominently displayed chartreuse candle. The heavy base completes the effect of a swirling triangle.

A piece of weathered wood surrounded by rocks and toadstools sets the pace for this one. A little beagle is tracking the elves who are hiding under the toadstool candles, and a big frog looks from behind a rock. Ferns and caladium leaves are natural companion plant material for the wood and rocks. Note the important part played by a proper base.

This fruit and vegetable arrangement is a joy to see. The gourds, tomatoes, and lemons are candles. A pineapple forms the backbone of the arrangement. Although most of the shapes are spherical, distinct contrasts in size and color provide interest, and in addition the cantaloupe was sliced in half to avoid monotony in form.

Croton leaves and ornamental gourd candles, both in green and yellow, combine with green zinnias to form a pleasing combination of related colors. This simple arrangement will last for days. The candles can be retrieved and used indefinitely, but are placed so they may be lighted.

A bird feeder makes an unusual holder for a candle. Pyracantha berries, a few pine cones, and poke berries form the natural habitat for the birds and create an artistic scene of nature in shades of red and green with accents of brown and black.

Natural plant material surrounds the little beehive candle, complete with honey flowing out of the hive. Mock orange blossoms and red clover blooms furnish food for the bees. The weathered wood base, which is adaptable to many designs, here gives stability. It is a diagonal slice from a seasoned hickory log.

Red-hot poker sets the pace. The colors of the flowers and container were chosen to emphasize the colors of the striped candle. The big hosta lily leaves balance the heavy candle, flowers, and container. The green and orange-brown stripes of the candle are produced by placing strips of colored wax into the mold before pouring the hot plain wax. The red-hot poker flowers were grown in my garden. The candle can be used; the plant material is not close enough to the candle to be singed.

I have a special pride in containers that I have fashioned from junk. (This one, which I use often, is described in the caption for the first photograph in the book.) The candle has orchid-colored cosmos blossoms molded on the inside. The same flowers are used in the arrangement. Broad variegated leaves add balance, and a bunch of grapes lightly sprayed orchid complete the color harmony. The candle can be used. As wax melts away from the flowers, the flowers are lifted out and the candle continues burning.

I was able to demonstrate flower arranging with candles using this arrangement, which won an award in a flower show. The container is a black wrought-iron candelabrum. The red flame candle emphasizes the rhythmic lines of wisteria, and calla lilies and their foliage further stress harmonizing forms.

Three stacked wooden bases form the container for this arrangement using the lovely "South Sea Miss" candle. The bold-colored Japanese peonies and huge green leaves suggest the colors of the islands. The wood tones coordinate the integral parts of the arrangement.

The black cypress base is the perfect container for the "Balinese Boy" candle. Orange red Maltese cross seed heads and blossoms are the perfect forms and colors of plant material for this little figurine. Groups of chartreuse sedum heads provide color contrast to both blossoms and candle. The satiny texture of the candle harmonizes with the fine texture and forms of the plant material.

The "Egyptian Goddess" candle, antiqued in gold, is used on a black styrofoam base. Lily leaves sprayed in antique green-gold, red gladioli blossoms, and buds with sheath stripped down give a zigzag line.

A purple dripped pillar candle is used in this arrangement. Spikes of purple foxglove blossoms complement the container, an exquisite antique crystal cake stand. Foliage of large leaves from gerbera and hosta lilies supplies contrast of line and color.

What wild flower but the dogwood brings the earliest news that spring is here? Blue candle dripped in wax, rubbed in white, forms the background. The wild sassafras blooms are the same shade of chartreuse as the center of the dogwood blossoms.

4

CHRISTMAS

Candles have traditionally been lighted not only for Christmas, and not only for other Christian holy days, but for holy days in many religions since the beginning of history. The lighted candle has different symbolism in different countries and cultures. But in this country, where candles are probably used more at Christmas than at any other time of year, they symbolize that Christ is the light of the world. There are also many special Christmastime candle traditions, such as the belief that a burning bayberry candle will bring good luck to the household.

At Christmastime, home is the best-loved place in the world. Christmas is a family affair, and each family has its own traditions of the Yule log, the Christmas tree, the Nativity scene, and Santa Claus. With a little imagination the old familiar things can be combined in a different design so that they appear new. Christmas is what you make it, so above all, when you decorate your home for Christmas, do it gaily!

The evergreens are always used in Christmas designs; they symbolize eternal life. "Holly" comes from the word "holy," and is symbolic of joy and peace. It is the most widely used foliage for Christmas. Sprays of holly covered with red berries and a red candle or two create a lovely Christmas arrangement to be enjoyed for days. The "St. Francis" figurine, representing the lover of animals and birds and the keeper of the garden, is widely used in Christmas arrangements. Usually the figure is in a brown or a green shade. The colors of nature need little to enhance them. Evergreen branches and figures of small deer or birds with a votive candle at the feet of the statue complete the design of quiet beauty at Christmas.

The dining-room table rates second only to the Christmas tree. Candles have a place of honor in table arrangements. Gleaming silver used with holly, red berries, and red candles are fine companions for red carnations, festive enough for any party.

Madonnas are a must as part of Christmas decorations, even Madonnas made from wax. (Of course, Madonnas are made from all types of material —wood, porcelain, glass, metal, and hand-blown glass.) The Madonna should be the dominant feature in any arrangement in which it is used; plant material should adorn the figure, not compete with it. Never light a Madonna candle; it is better to omit the wick when making one. Praying hands molded from wax complement a Madonna very well.

Driftwood combines well with pine, cedar, spruce, and almost any other evergreen. Add pine cones and figures of a few birds or small deer. This combination is a very good Christmas design for den or breakfast room.

A simple but unusual decoration that affords much pleasure is made by filling a brandy snifter, or any glass bowl, about three-fourths full of water, dyed as you wish with food coloring. Place a small floating candle on the water and light it. Another unusual candle for the Christmas season is made with salad oil, again in a brandy snifter or bowl three-fourths full of colored water. The salad oil is poured on the surface of the water to about a quarter of an inch in depth. Lengths of small candlewick (one-fourth to one-half inch) are threaded through holes in small cork disks. When these are floated and lighted they burn brightly on the surface, moving about as if by magic. The wicks absorb and burn the salad oil. Be sure the lower end of the wick does not penetrate the salad oil to the water beneath.

Christmas is the one time of year when we can let the imagination run wild. Use glitter, paint, and other bold and shiny materials. Evergreens can be lightly sprayed with artificial snow. Let your Christmas decorations reflect your own personality. By achieving this your home will reflect the warmth and love of the season.

All children love Christmas, especially the Santa Claus tradition. They should be allowed to help decorate. Variety stores offer a good selection of Christmas candles for children—little Santas, choirboys, angels, and so on. Children will also enjoy making their own candles, or decorating candles with glitter and tempera paints.

The following arrangements are particularly designed as Christmas decorations—but some of them would not be out of place at any time of year.

My black wrought-iron chandelier is converted into a hanging Christmas decoration, with Advent wreath attached underneath the fixture and red candles substituted for the light bulbs. Very attractive for a den.

The fireplace decorated to correspond with the chandelier shown in the preceding photograph. A piece of weathered wood is lightly flocked to resemble snow. Inside this wood is a miniature Nativity scene. Nothing shows the warmth of Christmas so much as an open fire and the evidence of abundance of the Della Robbia wreaths. Small twisted candles were used, and a red twisted candle completes the decorations.

The dining area of the den can be very effectively decorated like this. A straw wreath, colored with walnut spray, is used on the wall, decorated with a gold bow of brocaded velvet, orange bubble branches, brown wheat, and wooden fruit. This wreath would be very pretty on any door or paneled wall. The table arrangement is a Spanish wrought-iron candelabrum with artificial pine and holly and a twisted red candle. The red satin balls, in graduated sizes, are the finishing touch.

A beautiful piece of driftwood makes a magnificent arrangement suitable for a big foyer or the corner of a room. The holly and poinsettias suggest Christmas. The many bright shades of green, red, and orange are subtly unified by the Mexican pheasant, and the delicate white butterflies relieve the intensity of line and color.

Statues and candles go well together. Here small gold and orange twisted candles complement a brown Madonna outlined by a rattan bell and a collection of wild pine and holly. The color harmony is completed with orange bubble sprays (bare branches covered with small styrofoam bubbles—which you can buy or make yourself). The piñon cones add weight and a contrasting texture and form to complete the elements of this design.

Here the black wrought-iron candelabrum used in several other arrangements furnishes a playground for the mischievous little pixies. Red candles, poinsettias, and holly complement the candelabrum and pixies.

A large star candle forms the dominant element of this asymmetrically balanced design of white poinsettias, pine, and holly. The candle is decorated in one of the many possible ways; small gold angels add richness to the design.

A trio of pixies delight themselves at Christmastime. They are grouped around a black lamp-post candelabrum with snow-covered spruce and red Christmas balls suggesting the season. Like most arrangements that include fanciful figures, a delightful decoration for children.

A bright-red candle presides over this symmetrically balanced arrangement of Christmas holly, bubble branches, and deodara cones. The base is a discarded brass lamp base antiqued in white.

In an arrangement of two Christmas colors, dried palmetto palms sprayed chartreuse provide sharp contrast of color and form to the dominant twisted red candle on a red base. A bunch of red grapes adds richness to the dominant color.

The "Island Princess" candle in a Christmas arrangement. White poinsettias dominate the color and forms. Gold braid was used on styrofoam to add elegance.

A broken styrofoam ring sprayed red, a white deer, and red tapers make this unusual arrangement. The bubble pine branches add contrast of form and color. The strong vertical lines of the candles heighten the action of the reindeer leaping into the Christmas scene.

Blue pine was chosen for the chief plant material, and small blue candles strengthen the emphasis of the pine. The blues and the gold grapes supplement the regal splendor of the Madonna, antiqued in blue and gold.

A huge gold arrangement. All the material in the arrangement is glittered. The container is white antiqued in gold. The gold banana candle is glittered to harmonize with the fruit. A gold twisted candle blends with the design of the foliage.

Cardinals play havoc with the candle in this winter scene. White bubble branches contrast with the red candle and birds. The Swiss-cheese effect of the candle is achieved by filling the mold with ice cubes before pouring the wax.

This little angel is engrossed in a Christmas arrangement of red poinsettias, holly, and artificial pine. Her halo is a lace-paper doily. A candle is present not for decoration but for illumination—it is behind the little angel to provide a soft pattern of light and shadows.

5

OTHER HOLIDAYS AND SPECIAL OCCASIONS

Many holidays throughout the year can be the inspiration for arrangements. February has several. Candlemas, on February 2, is a little-known holiday, but it dates back to the eleventh century. Candles traditionally are blessed for church use on this day, and protection of the family from evil influences for the entire year is symbolized by burning tapers. The tapers can be used separately or in any arrangement of fresh, dried, or artificial materials.

Patriotic themes may be used for Abraham Lincoln's and George Washington's birthdays, using the red, white, and blue motif with or without flags. A small statue may be used in an arrangement. For rhythm use more white, less blue, and a much smaller proportion of red. Rhythm may be achieved by a large white candle as the dominant part of the color harmony, or tall thin red tapers as a subordinate part. Arrangements can be made using logs for Lincoln and cherry blossoms, hatchets, and cherries for Washington.

St. Valentine's Day. The arranger can also meet the challenge of February with cupids, laces, and hearts. Candles belong too. For St. Valentine's Day one can turn to the more sentimental arrangements, when we pay tribute to lovers young and old; there are the hearts made from various materials from velvet ribbon to wax. Small red hearts can be used in the center of the flower in red-and-white arrangements. "Hearts and flowers" can bloom abundantly.

Easter is one of our most beloved holidays—sharing with Christmas the religious theme, as well as the special symbols of Easter bunny and colored eggs. Christians the world over observe this day of Christ's resurrection. The Easter eggs came from an early belief in the spring goddess Eastre, who was thought to bring warm spring sunshine and flowers. The Easter bunny is a symbol of fertility. The Easter lily symbolizes purity. The Madonna, or the Virgin, is recognized by all Christian people as indirectly a symbol of Easter. All of these can be used effectively in Easter arrangements. Spring flowers emphasize the happy Easter holiday arrangement. The Easter eggs and bunny accessories will delight the youngsters. Baskets and driftwood are especially good Easter containers. Forsythia, Japanese quince, apple blossoms, and many other flowering branches may be cut while in bud and brought into the house, placed in water in a sunny window, and forced to bloom in about ten days. This will ensure flowers from your garden for Easter. Easter is the perfect time to have Easter-egg candles, bunny-rabbit candles, chick candles, and duck candles, for Easter arrangements as well as for the kiddies' Easter baskets. The little ones can enjoy such collections throughout the year. Den and breakfast rooms are the perfect places to play up design of the bunny and egg theme for the children, often using the Easter baskets for the centerpiece for Easter breakfast or perhaps a nest of paper straw and gaily colored eggs.

The Fourth of July is one of the holidays often observed with an outdoor picnic or patio party coupled with fireworks. For our Independence Day decorations, drums, flags, and firecrackers are appropriate. Use a drum for the container; combine flag and flowers to further emphasize the patriotic theme. As with other patriotic arrangements, red, white, and blue is an appropriate color scheme. In planning Fourth of July arrangements to use outside, take precautions against wind and sun; use plant material that has been well conditioned. Use hurricane candles or put ordinary candles in hurricane globes to protect them from wind.

Halloween is a fun time and a good occasion for a party. Halloween decorations can be as grotesque or as funny as we choose—pumpkins, black cats, goblins, witches, all are part of Halloween. Use a pumpkin for the centerpiece for the table. Use dried materials as well as fresh materials for Halloween designs—small candles in the shape of cats or owls add to the design and aid in strengthening the Halloween theme. Little witch candles can guard the punchbowl. Apples are also symbolic of Halloween. Float lighted apple candles in the punchbowl; this furnishes a good conversation piece.

Thanksgiving is the time of bountiful harvest. Play this up in your design for Thanksgiving parties. Fruits and vegetables combined with fall foliage and leaves are all that is necessary to have unusual and lovely Thanksgiving designs. Use the materials in different containers in a different manner—try small wicker baskets instead of the traditional cornucopia. Use candles abundantly, fruit and vegetable candles as well as other types. Make candles the dominant part of the design, with the plant material subordinate. In Thanksgiving designs make use of the harvest colors—dried corn, wheat straw, the beautiful colored leaves and seed pods, and the abundance of dried weed material that nature has furnished. Use them lavishly. The chrysanthemum is the most widely used flower for Thanksgiving. An elegant arrangement is made by using some candles anchored on a base or tray container with a few chrysanthemums and with many bunches of fresh or artificial grapes.

Weddings could hardly exist without candles—they are almost as much a necessity as the ceremony. Candles are often used more than flowers in weddings of good taste. A combination of the two is functional as well as furnishing beauty for the wedding and for parties. The designs for the bachelor party or the rehearsal dinner or party can be very bold and masculine. Candles should play a prominent part. Hurricane candles with flowers or fruits are very appropriate and beautiful. The groom's table at the reception should be very masculine too. For the bride's parties all the flowers and decorations should be very feminine and delicate, from the teas and luncheons through to the wedding reception.

Birthdays are the perfect excuse for an exciting party. Candles are always a must for parties young and old. A hobby can be the inspiration for the theme of the party. A large birthday candle that is burned each year until the child is sixteen or older can be the centerpiece; add some flowers and greenery, and for the younger child a few toys, for a decoration that will always be remembered. Candles can be made or purchased in a color or shape to emphasize the theme of a party for grownups. Some candles need little added to them to make an unusual design for a party.

There are numerous other holidays and occasions, not individually covered in this book, that are just as inspirational for candles and unusual decorations. The few ideas I have listed, and the color pictures that follow, are just to start you on your way.

The cupid candelabrum is a perfect holder for a valentine arrangement of fresh flowers. Spikes of white stock, white pompom chrysanthemums, and red carnations set off by leather leaf fern are the materials used in this arrangement for a bridal tea on St. Valentine's Day. Small hearts fashioned from chenille stems add the little touch that makes it different.

St. Valentine's Day is one holiday we all like to observe, especially children. This arrangement could be used for a party or just enjoyed by the family. A red styrofoam heart on a red base and red tulips are set off by white snapdragon, white foliage, and green tulip foliage. The arrangement is completed by using two little cupids, complete with bow and arrow.

Washington's Birthday—a modern version. A stump candle in brown, a hatchet made of wax, and the cherries make the meaning clear. The candle is the dominant part of this design; the red cherries and white blossoms add contrast and harmony of textures.

A colorful Easter arrangement. When lighted, the short candle at the base of the Madonna will produce interesting lights and shadows. Green palm or emerald foliage is clipped for more interesting lines. The red tulips give just the right color contrast, and a white leaf-shaped container completes the arrangement.

This little yellow bunny ran away to the woods. The stump he is hiding in is surrounded by pine, sassafras, and yellow tulips. A tiny green candle completes the Easter arrangement. Nice for a den or breakfast-room table

Easter in wax—bunny, eggs, chicken, and grape candles guarded by Mr. and Mrs. Bunny. These wax novelties can be used over and over again to go with each year's real eggs and candy bunnies.

Easter would not be complete without an arrangement of mixed spring flowers, Easter bunny, and eggs. The eggs on the left are wax, decorated in gold braid and medallions. The pink rabbit is also a candle.

The Fourth of July—red, white, and blue. A drum is used for the container. Blue lirope blooms, whitish-green caladium leaves, red peppers, and red geraniums furnish the patriotic colors of plant material. This arrangement could also be used for a child's birthday party.

A jack-o'-lantern is the natural container for a Halloween arrangement. Cattails, orange- and yellow-dyed pampas grass, and orange and yellow marigolds provide a natural setting for the little hoot-owl candle. The fresh flowers are arranged in water-soaked oasis.

A Thanksgiving arrangement, particularly good for the groom's table at a Thanksgiving wedding. The candleholders are made from old tablelegs. Orange and yellow fuji mums, salal leaves, and green grapes are the plant materials. The tall green pillar candles were decorated with dripped wax, then sprayed with pearl spray.

Thanksgiving—two small wicker wastebaskets are used in cornucopia fashion for this beautiful flower, fruit, and vegetable arrangement. Flowers, grapes, thistle, and pine cones spill out of the baskets. The potatoes, tomatoes, lemons, oranges, gourds, and grapes are wax candles made by Mary Wilson of Ohio.

Birthday parties are very exciting to children. Add the circus theme and the party is guaranteed to succeed. A square candle is used with a wax clown glued to the side.

Brandy snifters, stacked as in the last photograph in Chapter II, hold the wedding party. Lilies of the valley and white roses are the only flowers used. The bride and groom occupy the first tier, the best man and maid of honor the second, and the ring bearer and flower girl the third, all topped with a candle. The bride's bouquet, lying to the left of the arrangement, consists of white rosebuds, white camellia, lily of the valley, velvet grape leaves, and white pearlized grapes. I made this arrangement for my daughter in law's table at her reception.

This exquisite tiered antique candelabrum arrangement is designed for a tea table or a bride's table at the wedding breakfast. Tiny nosegay bouquets are alternated on the candleholders with candles encircled with greenery. The bouquets are miniature replicas of the bride's and bridesmaids' bouquets of roses, gysophila, satin streamers, and tuberoses.

Live grapevines, fresh grapes, and green pears were used to accent the black Spanish wrought-iron candelabrum. The small dripped candle in the hurricane globe was chosen with the colors of the plant material, so that in spite of its important position it does not dominate the arrangement.

State or local celebrations can inspire arrangements. This one, for an Alabama holiday, incorporates the Great Seal of the State, the State Flag, the State Tree (pine), and the State Flower (camellia). The pine is clipped to add distinction and make a more interesting arrangement. The State Flag and Great Seal are outlined in gold, then painted in colors.

6

CANDLES AND CANDLEMAKING

Candles found in retail stores today are usually just run-of-the-mill. They are round, square, or oval, with very few unusual ones. The dinner taper is about the only type to be found in most stores except at the holiday season, although some large department stores have a wide variety of tapers and twisted and plain candles of different sizes and colors. Seldom will you find one larger than an inch and a half in diameter. I do know of one company that has come out with a beautiful square taper. There is never any great variety of colors available in the ordinary market; one usually must place an order for any unusual color such as burgundy, avocado, or fuchsia.

During the Christmas season the supply of candles in the usual retail stores is expanded to include the large pillar candle in red, green, and white. Some department stores in the larger centers do have some of the larger pillar candles at other times, but few retail stores carry a good stock of candles the year round. Candles not sold during the holiday season are usually put on sale for quick disposition. Those remaining after the sale are relegated to an insignificant place in the gift department till the next annual interest in candles begins to glow.

However, it is not difficult to satisfy a year-round hunger for the artistic. There are many small secluded candle shops that take a real interest in their product, and some can make many special designs on order of the customer. For example, I have purchased candles in the form of statues, candles in the form of different fruits and vegetables, candles made from antique cut-glass molds, and many others. Most of these are interesting for themselves, but the attractiveness of many is their ability to fit into arrangements of flowers and into the artistic decor of any living or working area. These shops furnish not only the unusual candle, but also a wide variety of traditional candles in various colors not obtainable elsewhere.

The International Candle Crafters Guild is an organization especially for people who are interested in candles, both amateurs and professionals. It has regular annual meetings, and communication among the members is kept alive all year. Most of the members of this organization have candles for sale that cannot be found elsewhere. Some operate large businesses, and others make candles merely as a hobby. There are members in every state of the Union. Several of the members make their own molds, which gives them an almost unlimited variety; others confine their work to special types of candles.

Most candlemaking businesses are small, operated by one person, and therefore there is no production-line system and many do not have large stocks, but they will make candles to order. This should be kept in mind when corresponding with them and when inspecting their products. The beauty of a handmade product is often enhanced by small imperfections; a handmade article is often more valuable the more obvious its origin is. I must confess, however, that it is difficult even for the trained eye to find imperfections in some of the products of the members of this guild. A few businesses are listed below:

Candle-Glo, P. O. Box 3205, El Paso, Texas.

Decatur Florist and Nursery, Old Moulton Road, Route 5, Decatur, Alabama 35601.

Dillingham Candles, Santa Rosa Beach, Florida 32459.

Gladys McCain's Candles, 3005 McClure Lane, McKeesport, Pennsylvania 70130.

Glow Candle Company, 17 West 39th Street, Kansas City, Missouri 64111.

Lumi-Craft Candle Shop, Box 2, Norwich, Ohio 43767 (located on Route 4, East Zanesville).

Maggie Cripe's Candles, 338 N., East Street, Bremen, Indiana 46506.

Margaret Hosford's Candles, 4228 E. Easter Place, Littleton, Colorado 80102.

Myrtis' Candles, 1019 Danville Road, Decatur, Alabama 35601 (my own mail-order business).

Rene's Wax Creations, 606 Rue Royal, New Orleans, Louisiana 70130.

Twi-Lite Candles Incorporated, Lyons, Colorado 80540.

The following businesses offer beautiful stock candles:

Callaham's Floral Supply Company, 2130 Welton, Denver, Colorado 80205.
Colonial Candle Company, Hyannis Port, Massachusetts.
Emkay Candle Company, Syracuse, New York 13201.
Hall Palmer Florist Supply, 100 - 6th Avenue, S.W., Birmingham, Alabama.
Victrylite Candle Company, Box 890, Oshkosh, Wisconsin 54901.

MAKING YOUR OWN

The idea of making a "specialized product" in one's own home immediately discourages many people from entering into a delightful hobby. This is a pity, because candlemaking is not difficult, and the rewards are many. A little reading on the subject and a little practice, and you can make a large variety of candles of the specialized class, and you will not only have the satisfaction of producing beautiful handmade objects but you will have candles that cannot be purchased anywhere. Candlemaking affords much pleasure and relaxation; and it is an economical hobby—it can even become a business and earn you money.

Candles can be made simply by rolling sheets of beeswax around a wick, but the most popular handmade candles today are the molded ones. Proper equipment is desirable, and professional molds are usually made from metal or plastic, but any container that will hold hot wax can be used as a mold. The amateur can use water glasses, bottles, jello molds, cardboard containers, or many other items found in any kitchen. The top of a one-piece mold should be larger than the bottom to allow the candle to slip out easily after it has hardened. Professional metal molds are slightly tapered. Plastic molds usually come in two pieces so that they can be taken apart when the candle has hardened.

Most candle hobbyists start out making candles in their kitchen. It is well to cover the working areas with newspapers, over which is spread wax paper. This catches all spills and the wax may be remelted and used again. If the newspapers are not covered with wax paper, spilled wax will be discolored by the ink and cannot be salvaged. Do *not* pour candles near the kitchen sink, because hardened wax can clog the plumbing.

A metal pitcher is used for melting wax. The best method is to place the pitcher containing the pieces of wax in a pan of water; this is slower, but much

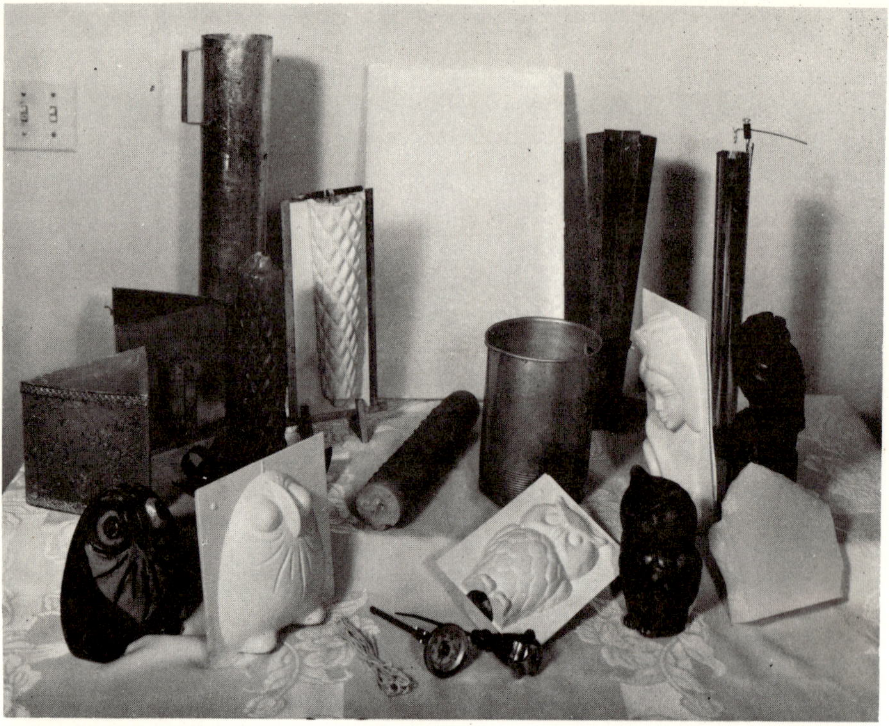

Candlemaking supplies. Plastic and metal molds, wicking, wax, melting pot, icepick for breaking up wax, and candy thermometer—and some finished molded candles. Note that two of the molds are strung with wick and ready for pouring.

safer, than direct heat. A candy thermometer is essential to make sure the wax is at the right temperature for whatever technique you are using.

Improvised containers such as milk cartons should be reinforced with masking tape at the top, the center, and the bottom to prevent a bulge in the mold and candle. The wick must be secured to the center of the bottom of the container with a professional wick tab or floral clay. The top end of the wick is tied to a pencil or small stick, which is laid across the top of the mold so that it holds the wick firmly, but not taut, over the center.

All metal molds have a small hole in the center of the bottom for the wick. The mold is threaded through the bottom and secured by a metal screw, wick

Candles. A very good example of what can be done with a little imagination, lots of wax, molds, and plenty of time. Most of these candles were used in arrangements described in this book.

tape, or clay. Each supply house has its own material for securing wicks. The wick at the top is secured as described for the cardboard carton, or with a regular wick holder that clamps onto the sides of the mold.

Plastic molds are usually in two pieces. They must be oiled on the inside with cooking oil, so that the candle will release easily when finished. The wick is secured to one piece of the mold, then the two halves are put together, like closing a book. The halves are secured by metal clips placed along the length of the mold on each side and on the bottom. Small racks are used to hold the mold upright. The metal clips and the racks are available at all supply houses. All figurine candles are made in plastic molds.

The wick is a very important part of the candle. The diameter of the candle determines the size of the wick. If the wick is too small it sinks into the melted wax at the top as the candle burns and the flame will finally go out. If the wick is too large the candle will smoke. There are two types of wick, the braided-cotton wick and the metal-core wick. The metal-core or wire-core wick produces a hotter flame, and thus the candle burns faster and is less apt to drip; therefore it is used in large pillar candles. Braided wick is easier for beginners to use, and professionals use it too for smaller candles.

The grade or melting point of the wax with which a candle is made is very important. Low-melting-point wax, which melts at 125° to 130° F., will not hold up under high summer temperatures or heating conditions. However, it is good for container candles. Wax with a melting point of 145° to 150° is desirable for most molded candles. A substance called steric acid should be added to the melted wax to harden it. This is added at the rate of about three teaspoonfuls per pound of good-quality wax that melts at 145° to 150°. More is required for the lower-melting-point wax. If the steric acid is purchased from a supply house, directions will usually be found on the container.

There are a number of ways to color candles. The regular candle dye available at supply houses is the best. This comes in the form of wax tabs, powder, or liquid. Some amateurs attempt to use wax crayons. This is not satisfactory, as the colors fade with the passing of time and crayons are expensive. Powdered Rit dye can be used fairly successfully. The color is stirred into the hot wax before the candle is poured. It is difficult to run a true test on the color in making a candle. Some candlemakers pour a small quantity of hot colored wax into a small bowl and cool it and test its color by eyesight. After some experience this is the most accurate method, but the finished candle is always darker than the sample. Only experience can teach one how to judge the coloring when a new shade is being sought. If an accurate record is kept of the amount of each color added to a certain quantity of wax the color can be reproduced with reasonable accuracy. Scent for scented candles may be added to the wax just before pouring into the mold.

Wax for pouring into metal molds should be heated in a metal pitcher to 190°. The temperature is measured with a candy thermometer, which is available at dime and department stores. Tilt the mold slightly and pour against the side to prevent bubbles from forming. Place the mold in a warm-water bath to three quarters of an inch from the top of the mold after the candle is poured. (A large plastic or metal container, such as a wastebasket, is handy.) This cooling method will give a very glossy finish. It will be necessary in most cases to weight the mold down by placing some heavy object on its base. After about forty-five minutes a well or indented place will appear in the center of the candle. Probe to the bottom of this well with a metal rod or a clean stick two or three times. Refill the well to the surface of the candle with hot wax. Continue this operation every forty-five minutes until the candle is cool and the well does not appear. When the candle is cool—this will take about eight hours for large candles—remove the screw from the bottom of the mold (if this method has been used to hold the wick). Turn the mold upside down and tap the bottom lightly. If the candle does not slip from the mold easily place it in a refrigerator for a minimum of one hour and try again. It is important to avoid heavy tapping on the mold, as the candle or the mold itself may be damaged.

Wax for pouring into plastic molds should be 150° to 160°. These molds are not cooled in water baths, but they do require refilling and probing when wells appear, just like metal molds. Filled plastic molds are placed in an upright position and allowed to cool at room temperature.

Floating candles can be made by pouring wax half an inch to a quarter of an inch thick into shallow molds, individual molds, or any small mold. Be sure the width or diameter of the mold is greater than the thickness or the candle will tip over. Metal-core or wire-core wicks are better in floating candles. These may be inserted as they are when using cardboard-carton molds, or they may merely be held in place until the wax cools sufficiently to hold them.

Any candle may be dipped in a hot-wax bath to glaze it and to remove any rough or defective spots. For this process the wax should be very hot, 240° by the candy thermometer, and deep enough to accommodate the candle. Slowly lower the candle while turning it; count to five at an ordinary rate, then remove the candle. If you leave the candle in the wax too long, the candle itself will start to melt.

To change the color of a candle, heat wax of the desired color to 240° and dip the candle into the wax, holding it by the wick at the top. Remove it after the count of five and wait about one minute for the wax to harden. Continue this process until the color desired is achieved. However, do not dip a sculptured figurine candle more than once or the delicate features will be damaged.

Almost any kind of bottle can be used as a mold, although there are special problems. An unusually thick bottle should be avoided, and it is much easier with a bottle that tapers evenly instead of bulging. Unless you saw the bottom off the bottle, it can be used only one time, for it must be destroyed in order to remove the candle. To proceed, oil the inside of the bottle thoroughly with cooking oil and secure the wick in a center position by weighting it down; this can be done by tying a dressmaker's weight, a metal washer, or any weight that will pass through the bottle's neck to the lower end of the wick. If the neck is large enough, position the wick in a piece of floral clay with a metal rod or clean stick. Tie the upper end of the wick around the middle of a match to hold it centered over the opening. The wax should be poured at a temperature of 160°. Then proceed as with any other candle, slowly plunging a wire or metal rod to the bottom and refilling the bottle until the candle no longer has a well at the top. Let the candle cool thoroughly before trying to remove the bottle. The big problem is the bottleneck, but if the candle neck is broken you can repair it with glue and hot wax. The bottle can be destroyed with a glass cutter or by wrapping it in a heavy towel and striking it sharply with a hammer. Strike the bottle on the bottom to avoid damaging the sides of the candle. If you are lucky, most of the bottle will crack and can be removed in pieces, and the rest will slide off, perhaps helped along by a little heat. Another method that may work if the glass is very thin is to put the bottle in the freezer for a couple of hours, then dip it in boiling water. Of course, if you can have the bottle's bottom removed by a professional glasscutter and if the bottle has the proper taper, you can use it like any mold, again and again. Position the wick by putting a piece of tape over half the opening in the neck, draping the wick outside the bottle, then capping or corking it. Invert the bottle in a jar, center the other end of the wick over the opening, and fill with wax at 185°. Whatever method you use, a quick dip in a hot wax bath will erase any minor blemishes caused by removing the mold. One decorating method is to drip various-colored waxes down the sides, giving the impression of a bottle that has been used as a candle holder.

Hurricane, or shell, candles are not true traditional candles as we think of them. They have no wick of their own. The flame is furnished by a small candle placed inside the hurricane candle. A hurricane candle is a shell of wax, open at the top. It protects the small candle within it from the wind; thus the name "hurricane candle." To make a hurricane candle, heat the wax to 160° and pour it into a hurricane mold—two triangular hurricane molds are shown in the illustration of equipment in this chapter. When the wax forms a scum on top about one quarter of an inch thick, cut the scum off the top, cutting around the wall of the candle, and pour the wax out, leaving only a shell of wax. Cool the candle before trying to remove it from the mold. The outside of these candles may be decorated in many ways. I melt leftover candles from weddings and make a hurricane candle for the bride, which is always appreciated. It makes a good keepsake for her, especially when some of her wedding pictures are placed on the sides of the candle as decorations. The candles may be decorated in many ways with braids, flowers, whipped wax, shells, glitter, beads, sequins, or anything you choose. The sky is the limit.

Figurine candles are especially suited to antiquing with metallic wax paste, which highlights the features. Paste is available at supply houses. Candles may be carved. Carve a niche in a large candle and use a small Nativity group or other three-dimensional scene that is available at dime and department stores. Frame the niche with braid. Carve all the way through a large candle and put in an angel, a deer, or some small object so it shows from either side.

Dripped wax is good for decorating pillar candles. Heat the wax to 120°, then let it cool until a skin begins to form. Tilt the candle to be decorated and pour small amounts straight down, diagonally or crisscross over the candle. The wax will harden in place. If the wax is too hot it will run all the way to the bottom of the candle; you will learn the correct temperature with practice. Rub with metallic wax paste to highlight the effect.

When your candle does not turn out as pretty as you think it should, camouflage the defects with decorations. Wax flowers and fruits can be applied to the candle with hot wax. Try to use your imagination and create some originals of your own. Add jewelry, braids, even pretty paper and pictures. If using pictures or paper, paint over with hot wax.

One of the most difficult types of candles is made with polyethylene flowers inside. To make these candles start by centering a core or a small taper inside the mold; secure this by pouring about half an inch of hot wax into the mold.

When the wax has hardened push the polyethylene flowers between the center candle and the wall of the mold. Pour the wax, heated to about 160°, into the mold. Process as you do any molded candle. The finished candle will show the flowers inside the candle, and the candle can be burned. Another unusual candle can be made by putting a taper in the center of a mold, as above, and filling it with small ice cubes. Pour wax, heated to about 200°, over the cubes. Your candle will have irregular holes throughout, resembling Swiss cheese. Such candles are very attractive. Fruit and vegetable candles are also very pretty. The color can be painted on with an ordinary paintbrush and very hot wax.

The materials and equipment I have mentioned are available at hobby-shops and can also be ordered by mail from some firms, a few of which I list below:

Candle Institute, 1600 Cabarillo Avenue, Torrence, California 90501.
General Supplies Company, P. O. Box 338, Fallbrook, California 92028.
Edith Miller Supplies, 416 Kalamath, Denver, Colorado 80220.
Lumi-Craft, P. O. Box 2, Norwich, Ohio 43767.
Pourette Manufacturing Company, 6818 Roosevelt Way, N.E., Seattle, Washington 98115.

Once you have learned the fundamentals of candlemaking—which will not take long—you will probably find yourself getting more and more ambitious, and you will want to take advantage of the equipment available at these supply houses and make advanced types of candles. You may even, like me, want to experiment with making your own molds, using a liquid molding compound and a special kind of plaster. Whatever happens, remember that in candlemaking there is no failure—you melt your failures and try again!